Towards a Magical Technology

TOWARDS A MAGICAL TECHNOLOGY

Tom Graves

Gateway Books, Bath
& Interbook Inc, San Leandro, Calif.

First published in 1986
by GATEWAY BOOKS
19 Circus Place
Bath, BA1 2PW

in the U.S.A.: INTERBOOK Inc
14895 E. 14th Street
San Leandro, CA 94577

© 1986 Tom Graves

No part of this book may be reproduced
in any form without permission from
the publisher, except for the
quotation of brief passages
in criticism

Set in Plantin and Kabel
by Wordsmiths of Street
Printed by W.B.C. Print of Bristol

British Library Cataloguing in Publication Data:
   Graves, Tom
     Towards a magical technology
     1. Technology
     I. Title
     600   T45

ISBN 0-946551-30-8

Contents

What's in a word?	1
Isn't it just coincidence?	3
Can't we explain this scientifically?	15
The subtle art of insanity	35
The practical art of magic	53
The joker in the pack	71
What's the use?	83
Bibliography	87

What's in a word?

People say sometimes that technology has lost its magic – its joy, its wonder, its meaning. But if technology has lost its magic, has magic lost its technology?

Magic, technology: two awkward words. But what's in the words themselves?

And what do we mean by magic and technology anyway? How do we define those words?

It's usually magic *or* technology: two separate worlds, and never the twain shall meet. One world is real, the other false: and your choice as to which is the 'true' one will say a lot about your point of view.

But the purpose of this study is to put the two words together: to look at them in an old way rather than a new one, a world in which magic and technology meet. In defining a word like 'magic' or 'technology', we limit its range of meaning; if we *un*define the words rather than define them – use them to describe rather than delimit – we open them to a wider context in which both words and their many meanings can work with each other. An exercise in thinking about thinking, and in exploring the often strange places to which this takes us.

In doing this, we can find a framework in which ideas and actions can be useful *in practice* – in which both magic and technology make sense in the real world we experience. The result should help us to put the magic back into technology, and the technology back into magic – moving towards a magical technology.

Is it just coincidence?

Magic is a peculiar thing. We know what we mean by the word – or do we? A dictionary defines it as 'illusion and trickery', but that's hardly what we mean when we talk of a magical occasion. To make practical use of a word, we need to have meanings from our own experience: magic as joy, as wonder, as wisdom. (Occultists talk of mages, while the biblical story refers to the three Magi – wise-men). Yet even these meanings are too limiting: we need words with meanings that somehow go beyond the words themselves.

Magic, many people would say, is just coincidence. Yet the word 'coincidence' is itself a prime example of the same problem. To say that magic is 'just coincidence' is to describe one known but undefinable 'something' in terms of another. For many people, each word is used to dismiss the other: neither is true, neither is valid, since, by definition, a coincidence is an event of no meaning. For others, every coincidence has meaning, part of pre-ordained fate or, for the paranoiac, a part of a deliberate plot against them and the world. A coincidence, it seems, has either no meaning at all, or far too much meaning.

But look again at the word. Co-incide-ence: two events coinciding, coming together in place, in time or in any other context. And that is all: the meaning, or lack of meaning, of the co-incidence is quite separate, to be derived from the context, the circumstances in which it occurs. If someone asks you whether some event 'was real or a coincidence', a proper answer would be "Yes": without knowing the circumstances, the total context in which to interpret it, there is nothing more we can say.

This is far from trivial, for everything we perceive is a coincidence of one kind or another. Our senses are geared to notice change, coincidence (and have real difficulty in 'perceiving' continuity). In fact, the *only* thing we perceive – whether sight, sound, smell, taste, touch, feeling or whatever – is coincidence, the co-incidence of events of one kind or another. What we think of as 'reality', the real world, is our interpretation of those coincidences.

We tend to think in terms of 'cause *or* coincidence'; but the idea of 'cause' is itself an interpretation of coincidences, so the choice of one *or* the other is hardly valid. For example, what caused you to be reading this book? You could say that the whole of your life has been a chain of coincidences leading to *this* word, at *this* moment. And any meaning, any 'fact', you may derive from these coincidences is your choice, your reality.

> Thirty spokes share the wheel's hub;
> It is the centre hole that makes it useful.
> Shape clay into a vessel;
> It is the space within that makes it useful.
> Cut doors and windows for a room;
> It is the holes which make it useful.
> Therefore profit comes from what is there;
> Usefulness from what is not there.
>
> Lao Tzu, Tao Te Ching

Events are what we see, 'what is there'; the meaning is 'what is not there', the context, the part from which usefulness comes. Anything and everything you realise – make real – has been and still is shaped by your interpretation of events, of coincidences. In effect, what we think of as fact – literally, a 'doing' – is our choice: we invent 'facts' to give meaning to those events. That's all there ever is: coincidence and its context, information and its interpretation.

What we think of as the facts of technology are better described as practical and predictable coincidences. And magic, as many people will tell you, is only coincidence. Just coincidence.

Working in the world

But to work on the world, we have to start somewhere. In this culture, it seems, most people would start by saying "Can't we explain this scientifically?" But this carries with it the assumption that everything is amenable to a scientific study – and this is itself a main component of our difficulties with working on the world, as we shall see later. Since we are dealing with *people* rather than abstract ideas, we need to start from where those people – you and me – are; we need to start with ways of working on the world, our skills at operating within the world.

Many skills we tend to ignore: walking, speaking, writing, reading and the like are all learned skills. Operating many kinds of machinery – a bicycle, a telephone, a car – is so much a part of us that we tend to forget how much we have learned, how many coincidences relating to them we have learned to interpret. For example, pressing one foot hard on the floor is hardly a natural response to danger, and yet that's what you learn to do when faced with danger in a car – and you'll find yourself doing it even if you're not driving! The 'explanation' of what you do and why says as much about you as it does about the processes you're involved in: in any skilled work, the same results may be achieved in totally different ways, according to how each person approaches the tasks in hand.

Learning a skill is a magical process, in many ways. A skill, in effect, is how each person resolves *for themselves* the mechanics of the skill – the 'real world' – with the way in which they approach it. Technology is what we see as the outward form of skills; and magic, perhaps, is the inner form. We'll be looking at this in more detail later on; for now, let's limit ourselves to the way in which coincidence and its interpretation form a key part of skills, and thus of technology as we see it in practice.

We should remember that we need to see this *in practice*: it's all too easy to drift off into a sterile discussion of theory without any practical grounding. So it's useful to keep some practical examples in mind. To keep up the idea of technology *or* magic for the moment, we'll use as examples, throughout this study, a skill from each side: the harsh world of computer programming, and the perhaps more dubious world of dowsing, or water-divining. What is interesting is that, by the time we've finished, you'll probably see that they turn out to be much the same: the way in which they work is surprisingly similar.

Let's start, then, with a (very) brief summary of what each skill involves.

Programming
For some people, the computer represents the ultimate de-humanising force in current technology; so it's worth remembering that a computer is only a tool. The English term 'computer' is confusing, since most computers – such as the one I'm using in writing this study – do very little computing, as number-crunching, at all; the French term 'ordinateur' – literally, 'an orderer' – describes the process much better. A computer is, in effect, a very fast but very stupid idiot: a logic machine, following logical rules or instructions, made up in a sequence that we call a 'program'.

Each set of rules is usually built up on top of other layers of rules: a programming language, then the 'operating system', machine-language below that, right down through the logic built into the processor to the simple logic of switches – two states, on or off.

Within the terms of the logic, everything a computer does is 'true'. But whether it is appropriate, or useful, is quite another matter.

But let's start by looking at the basic principles. And if we keep it to the bare essentials we can break it down to just five concepts.

The first is the idea of doing things *in sequence*. (Parallel processing, which is essential for handling anything happening at speed in the real world, handles many sequences simultaneously, passing results and instructions from one strand to another). One instruction is given, then another, then another – in some cases, millions of times a second.

The second key principle is the idea of naming things: 'if you don't know what it is, *give it a name*'. This is a key principle taken from algebra: that mysterious character called 'X', for example, who was *somewhere* between 1 and 3. (More accurately, 'X' *holds* a number between 1 and 3). Once named, this store or 'address' can hold some value, which might be constant or varied – as a 'variable' – by some other process, and can be referred to by its name. (Many programming languages, such as Pascal and C, need these names to be formally 'declared' within the program; other languages, such as most versions of BASIC, allow you to allocate names at whim; but the principle of something given an arbitrary name for practical convenience is common to all). The value stored in this named place is just a value: what it *means* – as a number, a letter, a place, a pointer, a reference or whatever – depends on the context: it's just information, waiting for the context to give it its meaning.

Which leads us to the third principle: an instruction can imply some *operation*, some context, using these stores by name. (An instruction is itself something that has been given a name; a programming 'language' is a convention describing a list of named instructions and their use). Most of these instructions are exceedingly trivial – it's only the sheer speed of processing that gives the computer program any semblance of intelligence. A typical operation might be to copy a value from one place to another, or to compare two values – almost the limit at the base-level of many processors – or, in a 'higher' level of logic, to show some value on the screen.

The next key point is the idea of *'perhaps'*: a logic decision, given in most programming languages as an 'IF... THEN... ELSE...' structure. '*If* so-and-so is true, *then* I'll do this, *else* I'll do that'. The context is built up logically, layer upon layer, with each decision made logically upon the context already defined by other instructions, other operations; the decisions are the 'intelligence' of the program.

Finally, we have the idea of *'do it again'*: usually after a logic decision, we can break the sequence, starting the sequence from another place, repeating instructions and skipping over others.

And really, that's all there is to it: layer upon layer of logic based on these five concepts. For example, here's a short program using terms from the programming language 'BASIC':

```
10 LET X=1
20 PRINT SPACE$(X) "Hello there!"
30 LET X=X+1
40 IF X<10 THEN GOTO 20
50 PRINT "That's a very basic piece of BASIC"
60 END
```

It doesn't matter if you haven't come across BASIC before: all that matters is that if you type this sequence of instructions on a suitable computer, and then type RUN, it produces this result on the screen:

```
 Hello there!
  Hello there!
   Hello there!
    Hello there!
     Hello there!
      Hello there!
       Hello there!
        Hello there!
         Hello there!
That's a very basic piece of BASIC
```

And yes, it *is* a trivial program.

You can see that we have a sequence: in this case, the line numbers, from 10 to 60. Each line consists of one or more named instructions.

Line 10 puts a number into a store which we label X.

In line 20 we use the current number in that store to tell the SPACE$() instruction how many spaces to print, followed by the words Hello there!.

In line 30 we add 1 to the number currently stored in X, and put the result back in X – so the number increases on each pass.

In line 40 we use that stored value to decide – depending on whether it is still less than 10 – to 'do it again' by going back to line 20 or to move on to line 50. Note that 'Hello there!' is only shown nine times – the question in the instruction is 'Is X *less than* 10?'.

And in line 60 we say that the sequence has ended, so that another layer of logic – the 'operating system' level – can take over.

Note how the meaning of the value stored in the place labelled X has different meanings at different points in the program: its meaning, its *use*, depends on the context in which it is used.

This example is trivial: it doesn't do anything useful. It's a set of made-up rules; the skill of the programmer is in making up sets of rules that *do* do something useful. As we'll see with dowsing, the question is not whether a program is 'true' – which it must be, by definition, since a digital computer of this type can only make true/false decisions of logic – but whether the rules it is given actually relate to the real world we experience. As with dowsing, the rules need to relate to the world in a way that is efficient, reliable, elegant (if you like), and, above all, appropriate: a way that relates to *people*. Otherwise, there's no point – an exercise in expensive and sometimes dangerous triviality.

Dowsing

You've possibly seen dowsers in action or played with it yourself. In principle, you wander around with some kind of instrument – traditionally a hazel twig, but nowadays more often a couple of bent wires or a bob on a thread – looking for something or other. When you stand over what you're looking for, the instrument reacts: the hazel rod bends up or down, the bent wires cross over, the plumb bob or 'pendulum' describes a circle hanging on its thread. In other words the reaction marks the coincidence of what you're looking for and where you are.

The reaction of the instrument is in fact its reaction to your hands moving; and your hands move because you tell them to. Not consciously, but as a reflex response to something or other – *un*defined. If you like, your hands move in response to a set of rules which you invent, which state that they *should* move when that coincidence occurs. It's much the same as with riding a bicycle: you *direct* the process rather than control it. Indeed, if you do try to control it by deliberate action, you're more likely to make a mess of it than if you leave it to 'work itself'.

In essence, that's all there is to it; which is why many people think that there *is* nothing to it. They can't understand how anything as crude could work, *therefore* it can't possibly work.

Which is to miss the whole point. Perhaps the shortest summary of the skill is to say that *dowsing is entirely coincidence and mostly imaginary*: and the catch with that, as we've seen, is that it depends on how you interpret those two words. Literally, the dowsing reaction marks the coincidence of what you're looking for and where you are; and both of those things – what you're looking for, and where you are – are defined by images, by descriptions, by imagination.

You decide what you are looking for, by describing it as an image. Traditionally, this was done by holding a sample of the sought-for material – a sample of water, or coal, perhaps, or some other mineral – beside the rod; but unless you insist that dowsing is the sensation of some as-yet-undefined 'radiations' (for which there is, of course, no 'scientific' evidence), there is no *structural* difference between holding a sample, and holding a written label, or a drawing, or even just a description of it in your mind's eye. Conceptually, they're all images.

Imagination is a powerful tool: we often forget how powerful. For example:

> Imagine that there's an orange on the table in front of you. (You choose what kind of orange it is: it's your choice, you're making up the rules here).
> You can see, in imagination, its colour and texture; see the way the light shows the dimples on its surface.
> Reach out and touch it; take some time to feel its surface, let the texture and weight describe itself to you.
> Now dig your fingernails in; feel and smell and sense the orange as the zest in the pith bursts out.
> Remove the skin, slowly, carefully; break the orange into its segments.
> Now put a segment into your mouth; feel the texture, then bite into it; taste the juice as it breaks through.

Now do the same with an imaginary sewer-pipe... you'll see just how powerful these images can be! It's all imaginary, and real at the same time – in an imaginary sense. And you can use that sense of reality to match what you're looking for; to note the coincidence between this image and the 'real' world.

You're also using images to describe where you are, where your 'current position' is. In basic dowsing, you 'are' where you stand: the rod (or whatever) moves when you stand over what you're looking for. Yet that's not all that practical when you're looking for something in a wall; so you change the rules and say that you 'are' where you are pointing to. Then you can change the rules again, and say that, having marked the point above what you're looking for, you'll now get another reaction at the same distance away from that point as the object is down: 'distance out equals distance down', sometimes referred to in dowsing circles as the 'Bishop's Rule'. Yet it's a made-up rule: it's a convenient image to describe something, rather than a 'fact' as such.

Follow the operational logic of that, and you'll see that there's no structural difference between someone out in a field dowsing with a hazel rod, and someone else looking for the same thing using a pendulum and pointing to various places on a map. A difference in degree, it's true, and probably a considerable difference in reliability; but no *structural* difference.

In both cases, people are following rules that they have invented, to pre-limit the meaning of the coincidence that the rod or pendulum marks. It's a significant point that dowsing instruments have a very limited range of responses: it's easier to pre-limit the range of possible 'answers' – coincidences – that way. In effect, you *declare* in dowsing that a coincidence shall have a particular meaning, and then set up conditions under which that coincidence can occur: if you like, you *program* the circumstances to have that particular meaning.

The catch comes in how well you can set up the conditions, so that a given coincidence *does* have the specific meaning you've declared. The point is not whether the reaction is 'true' – which it is, by definition, if you think about it – but whether it's *useful*: the key questions in dowsing are whether the method being used is efficient, reliable, elegant (if you like), and, perhaps most important, whether it is an appropriate tool for the job in hand. And the answers to these questions depend on the *people* involved, not on the outer form of the technology.

In both dowsing and programming, the rules we use are not pre-defined: we make them up. We *choose* to follow certain rules, for convenience and for useful convention to be able to discuss our results with others: but that is a choice, not a requirement. If you like, we use those rules to explain to others how we work. Yet within reasonable limits, *anything goes*.

But what are those 'reasonable' limits? And what do we mean by 'reason' anyway? It all depends, I suppose, on your point of view: it all depends on how you choose to explain things.

Can't we explain this scientifically?

Where once religion was the sole arbiter of Truth, science is seen by the main part of this culture as its leading light, the main source of its descriptions of the world. Science was (and still is, of course) a quest for knowledge in an abstract sense: but its public image, as presented in schools and colleges and in the media, is that of *the* way to analyse reality, using logic to define and delimit what is real and true (and what is not) in ever more minute detail.

In this context, technology is 'applied science'; religion is seen as an anachronism, poetry and the arts an irrelevance, while magic is nowhere to be seen – an aberration of the mind now finally eradicated by the ever-increasing progress of science in its explanations of the world and reality.

Our image of science is that it *explains* things for us. In fact, science is credited with all change, all advance, all progress. Which, as we shall see, is a little unfair: in reality, science has very little to do with it – and it has very little to do with science as practised. It's all a matter of your point of view.

What we call a point of view is better described as a 'filter', selecting out of the mass of information, of coincidences, those items that we consider to be valid, to be 'signal' events: everything else is just 'noise'. And yet we choose that filter, that definition of what is real and what is not. In the same way, we let it choose us: we can see only what the filter will let us see; a pair of rose-tinted glasses only shows us a rose-tinted world. If we cannot change the filter, we cannot change the way we see the world.

Yet we face a fundamental paradox. As Stan Gooch put it in a letter to the *New Scientist* magazine:

> Things have not only to be seen to be believed
> But also have to be believed to be seen.

Each point of view seems 'right'; the mistake, perhaps, is in assuming that if you are right, then by definition everyone else must be wrong. As Edward de Bono put it in his book *Practical Thinking*, 'everyone is always right, but no-one is ever right': they may be right from their own point of view, but they simply do not have the information available to be right in a total sense, to be truly 'objective'. And you can't really be objective, as most public descriptions of science claim to be, if all you have is a point of view. de Bono went on to describe two 'laws of thinking':

1 An idea can never make the best use of available information, because that information trickles in over a period of time.
2 Proof is often no more than a lack of imagination, in failing to see an alternative hypothesis that would equally fit the facts.

Or, as he put it in a rather less gentle form, 'certainty comes only from a feeble imagination.'

There are plenty of certainties around, pre-packaged ways of viewing the world: the catch is that the real world we experience never quite seems to fit these world-views. A political ideal like the beautiful communard slogan 'from each according to ability, to each according to his need', seems to come out in practice as 'from each according to facility, to each according to his greed'. And the religious world-views, for example, have a sad habit of assuming that people are 'perfect' – conforming to some arbitrary ideal – when they just aren't that way inclined.

Another example, the concept of the 'supernatural', is a crucial one to our study. The concept itself is another product of this limited thinking. The assumption that a scientific reductionism can explain everything also implies that the logical structure, the world-view, that it builds also defines and delimits the real world, the natural world. Anything outside of the structure is 'supernatural' – literally, above the natural – and, by a tortuous piece of circular reasoning, cannot exist: except, perhaps, as a product of the imagination.

Yet this distinction is quite artificial: if something is perceived, it's natural. It exists, even if 'only' in imagination, so it's natural. If everything is coincidence, and our perception of events is limited only by what we choose to perceive, then there can be no distinction between real *or* imaginary, natural *or* supernatural.

To work on the world *as it is experienced*, to work on it in different ways, we need to be able to explain the world in a variety of ways: we need to be able to change our point of view.

Points of view

In the last chapter we saw that coincidence and meaning are quite separate – we can't really say that anything or any co-incidence can be said to be caused by any other event, point of view or whatever. Anything goes.

The trouble with that concept is that if anything goes, we are left with nothing with which to make sense of the world. Without some way of handling it, we have no way of predicting anything at all, and everything and nothing happens at the same time – a certain recipe for instant insanity. So we need some kind of model which allows us the flexibility to allow anything to happen, yet still operate in something resembling a controlled manner.

One approach which is useful is to separate the information we experience from its interpretation, and describe this content and context each as an axis in a simple two-dimensional diagram. In one direction, we have a spectrum of information, ranging from outer tangible or sensory data, to information we derive from within ourselves, as feelings and intuitions. The other axis is a range of interpretation, from indefinable value judgements through to the strict true/false analyses of logic: a spectrum of interpretation between value and truth – whatever either of those might mean. The model looks something like this:

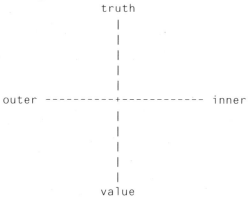

This gives us four quadrants, or modes, in which we both collect information and interpret it: inner value, inner truth, outer truth, outer value. Each is only a way of working on the world, a mode to describe reality as we experience it through that way of interpreting the world. Reality is, if you like, the sum of everything that could be experienced through these four very different ways of working on the world.

This kind of model of reality can be found in Jung's work, for example; but there is a particularly interesting variant on the theme in *SSOTBME*, a bizarre book on 'thinking about thinking'. The book was published anonymously so, for convenience, I shall refer to its author as Leo (a simpler name than his fictitious character Lemuel Johnstone).

Leo builds this model by describing the whole of reality as a swamp. Not a featureless swamp – every point of view, every experience, every possible coincidence of events is included. There are also endless opportunities to wallow in the mires of confusion, and to disappear beneath the surface without trace... It sometimes seems that the safest move would be not to move at all, to stay still. But even that isn't certain: the surface seems to quake with the tide of events, so that even the safest-seeming point of view will seem doubtful after a while. Nothing stays the same for long: indeed, the only real constant is change itself.

As Leo puts it, there are four main ways in which to exist within this kind of reality. Each one coincides with a quadrant of the model above: inner value, inner truth, outer truth, outer value. Each is best described as a *mode* of operation, in which certain possibilities – such as movement, in this sense of moving from one point of view or one experience to another – exist solely because others – such as stasis, developing one particular point of view – do not occur.

The first way of working on this world is to skim the surface of the swamp, travelling in a hydroplane at high speed. The whole point *is* the speed, and the variety of ideas and experiences that come from just travelling about with no particular place to go.

This is a mode of *inner value*, which we could call the artistic mode.

Playing with this description a little further, we can see that this is hardly a safe way of operating within a swamp: it's all too easy to crash into some unexpected experience, to run out of fuel (inspiration?) or decide to settle down in some uncharted spot with no hope of future supplies or common experience shared with anyone else. But it's certainly the quickest way of scanning a wide range of experiences and points of view – although, at that speed, it's not going to be too easy to make sense of anything other than that they were, indeed, experiences and points of view.

Another *modus operandi* seems quite the opposite of the artistic one: to develop one point of view as far as it will go, right out into another dimension. You state that that point of view is true – inviolably and absolutely true – and build on it, like a pole in the swamp.

This is a mode of conviction, of faith, of *inner truth* – the religious mode.

Again, playing with this image a little further, the higher you climb up the pole, the more of the swamp will come within your view: the more you climb, the more true will seem the point of view. In the distance you can see other poles, other points of view – some of them way out in the distance indeed – but you can hear that experiences from those poles, especially from further up each pole, seems much the same as your own. The mystics, those people who are well and truly up the pole and with their heads in the clouds, can see and share a vast range of vision – even though most of it seems like cloudy thinking to us.

The only trouble with this mode is that you can't actually *experience* anything else, since, by its definition, you have to stay with that one point of view; and it seems a sad fact that each pole has to be counterbalanced by a vast morass of struggling bodies, each of whom has grasped the pole and disappeared beneath the surface, screaming "I have the truth" as they did so.

> A friend and I were once intercepted at the station by an evangelist wanting to give us 'coffee and the word of God'. It was a predictable set-up, and the coffee was poor, too.
> After an hour of circular 'discussion', my friend yelled "This is a ****ing waste of time", and left – which was sensible, as we were about to miss the last train. I said "I'd better go" and made for the door. Our evangelist said "God Bless You" (I could hear the capitals); I stammered "Oh, G-god b-bless you". The reply came as I left the door: "But he already has..."

The third mode in this model is to build a solid platform, a safe predictable area in which everything is true and inter-related in logic. Everything is patently obvious, there are no surprises on the platform itself – although around the edges things may not be quite so predictable as they seem.

This is a world of *outer truth*, a scientific world.

To many people on the platform, the platform itself *defines* reality, and encompasses the whole of truth. To this point of view, which we could call public-science or 'scientism', anything beyond the platform is unreal; their duty is to build higher and higher walls around the platform, to protect the good citizens from the ignorance and superstition beyond. In fact this has very little to do with science as practised – we could suggest that these are same people who would have screamed "I have the truth" around the poles of religion, except that the solidity of the platform prevents them from decently disappearing beneath the surface as would have happened elsewhere.

The platform is woven between a group of poles, more often called the ivory towers of academia; their mystics are the 'pure scientists' whose breadth of vision is matched only by the impenetrability of their thought. And at the edge of the platform are practical scientists, researchers working at the limits of the known world (having found the occasional hole in the fortress walls of scientism).

Scientists, says Leo, are like people in wheelchairs – they have to have firm level ground to move about on. To move, they must extend the platform, extending the boundaries of science, cutting down shady dogmas and filling in soggy hypotheses (to use Leo's graphic image). But when they arrive at some new place, it is just as boring and predictable as anywhere else on the platform – hobgoblins and foul fiends having rushed away at the sound of the first myth being exploded. For the problem with this mode of working is that, by the time it has finished, what it seeks has ceased to be the swamp, has ceased to be reality as it is (or was) – it is just an artificial platform, an 'objective' world with no room for personal experience at all.

But working away at the edges of the platform are another group, commonly but quite wrongly called 'applied scientists'. At one edge you'll find the psychiatrists, not bothering too much about which theory is absolutely true, but using ideas from Jung, Adler, Freud, Laing and anyone else's work they can lay their hands on. At another edge there'll be electrical engineers switching between wave and particle theories of light and energy, blithely unconcerned about their mutual incompatibility in logic.

This looks like science, but only because of the safety-line of 'if it doesn't work, go back to theory' – in other words go back to 'outer truth'. But in fact this is a quite different mode, in which you carry the platform with you, spreading your weight on swampshoes to allow you to move with relative freedom from place to place, idea to idea, to find a point of view which is *useful* rather than necessarily true at that time.

This is a mode in which truth is defined in terms of whether it has practical value, *outer value*.

We could describe this as a technological mode. But it has an easier label – a magical mode. *There is no structural difference between magic and technology.*

> "Any sufficiently advanced technology is indistinguishable from magic"
>
> Arthur C. Clarke

The only difference between magic and technology, in practice, is that magicians tend to be a little way out in the distance – they may be seen wandering to astrology, alchemy and other forgotten, part-ruined areas of the old platforms of science, to rest by some religious pole, or travel as fools where angels fear to tread.

This mode is hardly *safe*, as the platform of science may be, but at least it works on the swamp as it is; and the point of the exploring is not to find out how true a belief, a point of view, may be, but to put it to *use*.

Technology is not 'applied science'
If we use this model, we can see clearly that technology is not 'applied science'. In fact, it's quite the other way round: in our culture, science is a codified summary of the practical experience of technology, simplified to describe what would happen under nonexistent 'perfect' conditions. Reality is always a little different from the niceties of theory: for example, the pressure-temperature-volume equations, the 'gas laws' we learnt at school, only apply to a perfect gas that doesn't actually exist.

One of the biggest problems in technology is not the neat precision of theory, but of getting things to work in the real world. Leo, the writer of *SSOTBME*, said that one of his problems working as a mathematician in aircraft design was to make his maths sufficiently *im*precise to be useful. An important recent advance in computing practice has been the development of 'fuzzy logic' – the mathematics of being precisely imprecise.

Varying tolerances in materials can make a mockery of any design, too: low-cost versions of electronic components, for example, may be up to twenty percent away from nominal values – multiply that a few times and things can go a long way from the neat predictions of the textbook. One of the real arts of electronic design is in building circuits that cancel out any problems from loose tolerances and lower-cost components. There's a popular truism, too, that defines an engineer – a technologist – as 'someone who can make something well for a pound that any fool can make badly for ten'. In that sense, technology is an art-form in its own right.

And so it goes on. Historically, too, most technology comes *before* science: intercontinental radio and heavier-than-air aircraft were considered impossible until theory caught up with practice. The idea of scientific research actually driving progress is mostly a myth: far more often it's been someone chasing an idea, or looking for commercial advantage. Technology does the work, but technologists call themselves 'applied scientists' to improve their credibility rating in the academic sweepstakes; science gets the credit.

Science is presented as the controller of a linear progress from idea to hypothesis to theory to law. But we can see that this is only another way of looking at the same diagram:

```
idea  -->  hypothesis  -->  theory  -->  law
```

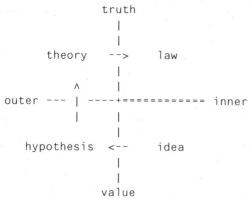

except that there is a rigid wall between the finality of 'scientific law' – 'inner truth' – and the chance of any new ideas. Historically, science has progressed as a sequence of 'scientific revolutions', to use Thomas Kuhn's phrase: the Copernican revolution, the Newtonian, relativistic mechanics and so on. An idea is put into practice as technology, is codified as theory, and frozen as scientific law: and the rigidity and finality of law prevents us from seeing the *context* in which that 'law' is true. The 'law' is how reality *ought* to be if the world was perfect: the law tells us God's intentions, as the religious person would put it. Or perhaps it's us telling some God how the world ought to have been made...?

Explanations, explanations

What it's really about is *explanations*. An explanation is a model, a prediction of how something will work. Taken in a wider context, a group of explanations define how the world 'really works': and if something doesn't fit the explanation, *it doesn't exist*.

The explanation is seen as being more important than reality, the totality of the swamp. Because the explanation *predicts*, it makes the world *safe*. Or seem safe, at any rate. Everything is known, everything is predictable, there is no supernatural, 'I am the Way, the Truth and the Life, there is no way to the Father but by Me', 'there is no God but God, and Mohammed is his prophet...' – which can lead to a rigid state that Edward de Bono described as 'nothing-but-ness'.

A central facet of this kind of closed explanation is the ascription of one unknown to another – a tautology – to prevent any questioning of the belief-system. "There are only four forces in the universe – weak nuclear, strong nuclear, electromagnetic and gravitational; and everything is causally connected", wrote the popular-science writer and pseudo-scientist John Taylor, neatly sidestepping the fact that we do not know what any of these forces are, and have no causal explanations for even key concepts such as radioactive decay.

An explanation is used to 'explain away' anything which doesn't fit. Armchair Freudians use a system so neatly closed that if you say you don't need psycho-analytic treatment, that statement is proof that you need it! And I've had many a battle with the 'religious left' who believe so strongly in democracy that, as the only true representatives of 'the people', only people who agree with them may be allowed to speak or vote – anyone else is, by definition, 'an enemy of the people'.

A few years ago, ten young Germans became convinced that their country was a rigid police state. The group, led by Andreas Baader and Ulricke Meinhof, conducted bombings and 'executions in the name of the people', to fight the cruelty and injustice of the state as they saw it.

The state's response was that all police were mobilised, road blocks and 'search-on-suspicion' orders were imposed throughout the country for years, and civil liberties and legitimate forms of protest were severely curtailed: the characteristics of a police state.

Germany hadn't actually been much of a police state when the Baader-Meinhof group started; but it certainly was when they had finished.

Their belief system hunted out the 'facts' it needed to 'prove' that its assumptions were true at the end; if the facts didn't exist, *it created them*. It had no means of testing whether its assumptions were *valid* in the first place, or whether the actions dictated by those beliefs were *appropriate* at all.

When explanations are used in this way, only those things predicted by the belief-system can be seen to happen: which is, of course, the whole point of using an explanation in this way. Remember that 'things have not only to be seen to be believed, but also have to be believed to be seen': so the assumptions about what will happen are also the cause of what does actually happen.

The danger is that while theorists and politicians play at 'what I tell you three times is true' (to quote Lewis Carroll), the real world isn't quite so accommodating. Whilst change is not always for the better, a rigid world-view that cannot cope with change at all will, sooner or later, sink into the swamp, often dragging others with it into insanity beneath the surface.

> The writer Dr Johnson was once standing in a street where the houses came closer and closer as they reached towards the sky. He saw two women hurling abuse and vegetables at each other, almost within frying-pan range from the topmost floor of two houses on opposite sides of the street.
> "They can never agree", said the good doctor, "for they are arguing from different premises."

An explanation is a tool, not a 'fact'; it's a way of describing what was perceived. Like technology itself – 'the use of tools and techniques' – we can either use them or, through ignorance, be used by them. It's our choice.

Where people believe that their safety and sanity depend on a belief structure, an explanation-system, that structure is more important than themselves. They no longer direct it: *it* controls *them*.

The logic, 'the Truth', that they perceive tying the structure together is like a single thread holding together that platform of science – 'the real world' – linking everything in one unified structure which defines their world-view, their definition of what is 'sane' (literally, 'healthy') and what is not. Break the thread in just one place, though, and the whole complex edifice collapses... and with it their sanity. So it's no wonder people will die for a belief, will go to war over beliefs: structures like this are terrifying fragile.

The world-view becomes rigid, and the social definition of sanity is itself insane: a mad world in which security can only be achieved by the threat of 'mutual assured destruction'. To protect the strand of logic, 'reality' can only be seen in terms of the structure, is in fact *defined* and delimited in terms of the structure. And we have a tautology: the structure is 'proved' by logic to be true in terms of itself, from its own viewpoint, in order to prove that the logic was, indeed, logical.

But we knew that already: what is the use?

'Use' comes from the context, from 'what is not there'. No doubt you remember, from your school-days, that the sum of angles within a straight-sided triangle is 180°. It's part of the logic of Euclidean geometry. But this is only true for a flat surface, not a spherical one, as many sailors have found to their cost: on a sphere, 'straight' lines are curved into another dimension, and the sum of the enclosed angles may be anything up to 540°; and in other geometries the values are even more bizarre. The truth, the 'right' answer, changes according to the context.

Yet how do you tell, sailing on the sea, whether the world is flat or curved? And if curved, whether it is spherical, or cylindrical, or sausage-shaped, or anything else? Common sense, after all, says that it is flat; you need a context greater than your immediate perception to understand the idea that the world we stand or sail on is curved. And even that is an assumption – a *useful* point of view. In any technology that deals with the world as it is, rather than as some theory pre-defines it, anything may be 'true': what matters is its *usefulness* – whether it is efficient, reliable, elegant and apt.

> "What is matter? And does it?"
> "Now this", he said, raising his glass, "is a very different matter. And it does!"
>
> advert for *Guinness* stout, c.1970

Where beliefs and points of view are seen as tools rather than final Truths, a world-view is held together by many strands: some may break from changing circumstances, or simply from old age, but the structure is flexible enough to withstand it. A tatty collection of old ropes of thought may not look so neat as a so-elegant single strand of logic, but it least it isn't fragile; the structure may sway a little in the winds of change, but it's not so likely to collapse without warning in the minor earthquake of some 'scientific revolution'.

So, to work on the world as it is, rather than on how some belief defines it to be, we need some way of seeing the context wide enough to select an appropriate point of view. We need to move from explanations that *define*, to explanations that *describe*; we need to move from point to point within the swamp, yet have some way of knowing where we are.

We call this process 'learning'. And yet, since to move from belief to belief is unstable, dangerous, we also call it insane. Learning a skill, learning to work with the world, could also be called 'the subtle art of insanity' – and that's what we'll look at next.

The subtle art of insanity

Most of us are afraid of insanity, yet in a sense we go insane every day. Being sane is keeping your thinking on a pre-defined track, keeping on some pre-defined train of thought.
The social definition of sanity, the 'clean' way of living, the 'right' track, is only one definition, one world-view: indeed, it could be argued that the current social definition of what is sane is itself so insane – in the literal sense of 'unhealthy' – that we cannot be personally sane in this culture. To be personally sane, it seems, we must be socially insane.

We have arrogant images of other cultures, but fail to see the insanity in our own: we can't see our own point of view from that point of view. For example, many people express horror at the enforced placement of Soviet dissidents in mental hospitals: but look at it from the Soviet point of view. From that view-point the Soviet state is, by definition, the nearest to perfect that can be achieved, so anyone who disagrees with it is mad, and should be placed in a mental institution for their own good. And we do much the same, in our own ways: anyone who doesn't fit the neat rules of the system – like the eccentric and the crank who moves in different circles – is neatly forgotten, or hounded as a 'subversive'.

To admit such insanity – anything which doesn't fit – into the sane world with its so-sane, so-safe belief system is indeed subversive: anything which can shift our way of thinking from its pre-defined tracks could destroy the whole tidy but brittle logical structure, and with it that sense of reasonableness, of 'rightness' – 'with God on our side' – that that structure brings.

Yet that is exactly what we have to do when we learn any skill: we have to change the rules, the assumptions we use in working on the world. Looking at a bicycle for the first time, you *know* you cannot ride it: with only two wheels, it's inherently unstable. In particular, you know that *you* can't ride it. Yet all around you are people riding bicycles... So how do you learn to include into your world-view the new rules necessary to ride it; and what would the use be, anyway? What's the point?

In essence, how do you learn how to discover; how do you learn how to learn? Not, it seems, by being 'reasonable':

> The origin of discoveries is beyond the reach of reason. The rôle of reason in research is not hitting on discoveries — either factual or theoretical — but verifying, interpreting and developing them and building a general theoretical scheme. Most biological "facts" and theories are only true under certain conditions and our knowledge is so incomplete that at best we can only reason on probabilities and possibilities.
>
> W.I.B. Beveridge, *The Art of Scientific Investigation*

Being reasonable can only tell us more about what we already know, in a logic we already know: it cannot tell us or give us anything new other than this sense of 'rightness'. If we are to learn, we must discover — and this is, as Beveridge says, 'beyond the reach of reason'.

This analytical mode of thinking, on which most training is based, works only with the statement "Here is a question: what is the answer?". It can only be used to *deduce* an answer from premises, assumptions, already provided, where all the conditions are known: it's not so successful in the real world, where 'facts and theories are only true under certain conditions'. In practice, real-world problems tend to come up not as the tidy "Here is a question: what is the answer?", but something more like "Here is an answer: so what was the question?"

I know this problem only too well from computer programming. A digital computer, by definition, is a logical beast: unless something is physically wrong with its electronics, it will always do exactly what it is told to do, and *only* what it is told to do. So when a program 'goes wrong' – does not produce the results I wanted – it's not the machine's fault: it did exactly what I told it to do, it answered precisely the logic I gave it. So what on earth did I tell it to do?

It's relatively easy to analyse your way through a simple program, but the process can be slow; for a complex program, ridiculously so, to the extent that proper logical test procedures for many programs could take thousands of years to run. So if we can't be logical, we have to be illogical – or at least non-logical – in our thinking.

In debugging a program, everyone's techniques will be different: there is no set method. If you rush in with assumptions about where the problem is, you can easily spend hours looking in entirely the wrong place: it pays to be slow, methodical, careful, listening to the 'sense' of the problem and the conditions around it. One technique I tend to use is to look at things from the program's point of view and, by pretending to 'be' the program, get the program to tell *me* the circumstances, the context, in which the error occurred. But anything goes, anything will do; the only criteria, as we've seen before, is that it should be efficient, reliable, elegant (if you like), and appropriate.

Note, too, that a technique that's appropriate at one point – checking the values of variables, for example – may be a waste of time at another. You build a 'toolkit' of ideas, of techniques, of methods, using each as appropriate. What you do *not* do is stick slavishly to one technique: that's sometimes known as the 'Birmingham screwdriver' approach – using a big hammer for everything! It has some unexpected uses, perhaps, but not for *everything*...

The failure of logic

Yet this is only within the artificial world of a computer program: all the rules have been pre-defined and are, somewhere, known. Many people would, I suppose, say that the aim of science is to discover all the rules of reality, the 'laws of the universe', and link them into one tight network of logical correlations: but this is, in reality, doomed to failure by its own logic.

At first sight, a logical analysis of reality, a reduction into ever smaller and smaller parts, looks as though it would actually work: but in practice this approach is a classic example of what de Bono describes as the 'magnitude mistake'. Reality is simply too big. To enmesh the whole of reality into this network, we must include every factor, every incident, every possibility: if we fail to do so, we cannot reliably analyse *anything*, since it may be affected by some factor we have not yet included. For each factor we tie into the net, we increase the correlations factorially; when only two factors are compared, there are only two correlations, with three there are six ($1 \times 2 \times 3$), for four there are twenty-four ($1 \times 2 \times 3 \times 4$), and so on. With only seven factors, seven aspects whose variations have to be taken into account, there are over five thousand correlations to make; yet in our experience, we know that there must be an almost infinite number of factors to take into account whenever we work on the real world.

So we have an impossibly huge number of correlations to make – the nonsense number of 'infinity factorial' – which, since each correlation takes a finite amount of time, however it is done, will take an almost infinite amount of time to do. And since many of the factors operate within time, and are related as events within time, we have an impossibility: not in theory, but *in practice*. We cannot do it: as is so often the case, science works as a world-view in theory, but not in practice.

If we are to work on the real world, in practice, we simply do not have the time to analyse everything. In practice, to learn about how people learn how to learn, we must look elsewhere for guidance.

'Beginner's luck'

The sequence with which we seem to learn many skills gives us a good starting point. At first, we can get the right result straight away: 'beginner's luck', we call it. Then, after that initial burst of success, everything seems to collapse: we cannot hit the target however hard we try. So after a long period of practice, confidence and skill return: we finally know what we are doing, and how and why.

> We shall not cease from exploration
> And the end of all our exploring
> Will be to arrive where we started
> And know the place for the first time
>
> TS Eliot, *Four Quartets*

But let's look at this in a different way. At that beginner's stage we know what the target is: so we get there. Without preconceptions, without assumptions, and with enough encouragement to convince us that it is possible, we simply get there. For example, on occasion I've had to teach dowsing to people very quickly: one example was on a television chat-show, where I had less than three minutes to show ten people how to find, as a team, a pipe under the lawn outside the studio – live, in front of the cameras. They did it, too. Whether they could do it again after the show was another matter...

The crucial point is that when we first start at a skill we know the target, but we do not know 'the way', *the* method, that we are expected – by tradition, by 'experts' or whatever – to use in order to reach the target: so we are free to invent any way of our own that will get us there, reliably. We make up the rules as we go along, in whatever way suits us.

The catch is that we often don't know what rules we've invented: we then have a long haul through analysis of one kind or another, through other people telling us 'what we really did' – and which usually stops us from getting anywhere near the target for a long time. It's only when you find out, through practice, what works *for you* that you can start again, working towards the target: and this time, since you've studied all the factors, in practical experience, you'll know how it works – for you – and why it works – for you. But not necessarily how and why it works for anyone else. You're following rules that you have invented for you: it's your point of view, not 'fact' at all.

There are facts, of course: the materials you work with, their responses, their textures, their strengths and weaknesses. You learn, quickly, that you don't apply the same force to a glass screw-thread as you would to a brass one. Even the most rigorously selected materials will have variations, which you will have to allow for; you may have to change your design, for example, to go round a knot in a piece of wood. The study of these problems, in all their fine detail, makes up what most people think of as technology.

But perhaps the most important facts relate to the way in which *we* approach those facts of the physical world: the study of 'gumption', that uncommon 'common sense', as Pirsig described it in *Zen and the Art of Motorcycle Maintenance*. Or, as Beveridge put it in his Preface:

> Elaborate apparatus plays an important part in the science of today, but I sometimes wonder if we are not inclined to forget that the most important instrument in research must always be the mind of man. It is true that much time and effort is devoted to training and equipping the scientist's mind, but little attention is paid to the technicalities of making the best use of it.
>
> W.I.B. Beveridge, *The Art of Scientific Investigation*

To put it another way, we need to develop a technology of mind.

And that is exactly what many people would describe as magic. This is more than that sense of magic that comes through in the process of discovery: the magical tradition has developed as a group of technologies in its own right. Historically, those technologies are primarily concerned with 'the technicalities of making the best use of [the mind]' – and through that, to changing the world through the way that we see the world and ourselves.

To start again from a different viewpoint, that of the psychology of skills: We could say that skills, as experienced, can be separated into three interdependent aspects: the *mechanics* of each skill, the outward form and the physical principles involved; the *methods* and techniques used, the outer actions; and the *approaches* of the craftsman, the operator to the work being done.

To say the same in a different vein, the mechanics of the skill are the *objective* aspects, those which are common to everyone; the approaches to it are the *subjective* aspects; and any methods used are the means by which the craftsman, the operator resolves those aspects of themselves and the 'real' world in realising – literally, making real – the skill.

In effect, a method is a technique is also a belief. You make it up, you invent it: within the reasonable limits of the mechanics of the skill, anything goes.

The objective and the subjective – the mechanics of the skill and our approaches to it – define the *context* of the skill. You derive meaning yourself from that context, from the skill as you perceive it and from what you believe about it and yourself; and put it into practice as actions, as methods, as techniques.

So it's interesting to note that in present-day skills education, almost all training is based on method – which, as we can see, is the worst thing we could do. Very occasionally – as with Philip Harben's books on cookery, for example – we see the mechanics of the problems discussed in detail before methods are presented: but it's definitely unusual. And references to the 'approaches' aspects are extremely rare: the classic is Eugen Herrigel's seminal *Zen and the Art of Archery* – from which Pirsig derived his whimsical title of *Zen and the Art of Motorcycle Maintenance* (followed by a few Californian-style titles like *Spiritual Tennis*, which I'd prefer to forget about). It was precisely because there was no book available for his students that Beveridge wrote *The Art of Scientific Investigation* – and, thirty years later, it's still almost the only book of its type. And so it goes on: a typical problem created by by people trying to tackle technology as if it were 'applied science'.

On occasion, when technology has been viewed as practice rather than theory, there have been books about its practical realities. One that comes to mind is *How to Invent* by Laithwaite and Thring (pioneers in magnetic-levitation transport and magneto-hydro-dynamic electricity generation respectively), in which they stress the importance of 'thinking with the hands' and using chance, imagination and analogy as sources of ideas. But these are still so depressingly rare: as Pirsig put it, perhaps what we really need in universities are courses on 'gumptionology', the study of common sense...

Inventing reality
But let's return to the magical side of the discussion. Once we understand that paradox 'things have not only to be seen to be believed, but also have to be believed to be seen', we can start to understand at least one peculiar aspect of skills. We have to believe that we can do it: otherwise our tramlined thinking, following its usual trains of thought, cannot conceive of the possibility of our doing it. We can't, for example, simply try harder:

> Relax.
> Try harder to relax.
> Not good enough – try harder!
> Now you must put some *effort* into this: *try hard* to relax.

which, unless you're one of the lucky ones, will have left you anything but relaxed...

We have to *invent* the possibility of doing something new, we have to create this from nothing. Indeed, as James Burke suggested in his book *The Day the Universe Changed* – a study of the cultural impact of changing technology – we don't really discover new 'facts' about reality: we invent them.

This has some important consequences. One side of this I first came across in a science-fiction story called *Noise Level* (I think it was in a collection called *Orbit 5*, but I've long since lost the book). In the story, a group of researchers are called to a conference, in which they are shown convincing filmed evidence of an anti-gravity device. The inventor, now dead, worked alone, in a house stuffed full of all sorts of confusing items: a fully equipped engineering workshop and a physics lab, for example, side-by-side with a very complete library of books on magic.

Eventually, by working through the concepts involved, the researchers in the story re-create a just-about working anti-gravity device. Or think that they've re-created it: in fact they've invented it, for it turns out that the original device was a fake, the 'dead' inventor was an actor, and the whole conference an exercise in thinking – giving them enough random information to open up the 'noise filters' of their previous assumptions and let them invent new ways of working.

We invent new ways of working on the world every day. For example, we all know about 'willing' traffic lights to change, or 'manifesting' a parking space just when there's no parking to be had. Or, as another friend described, learning how to sense when his motorcycle needed attention, rather than relying on textbook checks and timetables. All coincidence, you could say, of course – but putting coincidence to *use*.

Another side of this takes us even further, beyond 'coincidence' into areas that are more readily thought of as magic. Here's one first-hand description of an extraordinary incident in otherwise very ordinary circumstances:

> It was late one night, and a member of our small party [of musicians] had left his bags in the studio reception area. Unfortunately, on arriving at the studio, we found the door locked. We rang the bell, and waited [for the security guard]. We rang several times, but nothing happened.
> Finally, one of our group walked forward, inscribed some brief designs on the door with his finger, and mumbled a few words under his breath. With a loud bang, the door swung almost wide open. I reckon it would have taken the simultaneous impact of three people on the door to have the effect of opening it so far. We went in before the door closed on its spring, and went over to the bags.
> The door didn't close completely, however: it couldn't, as the tongue of the lock was still in the locked position. It was as if the door had been unlocked, pushed open, and relocked in the open position.
> But this had not occurred... neither had I been hypnotised, and so persuaded that the door had been opened by an invisible agency: that would not have got me past the door, and it would not have explained the bewilderment of the security guard when he finally arrived. What *had* occurred was a magical operation with an objective result.
>
> Richard Elen (unpublished manuscript)

(Since including a fuller version of this report in a previous book, *Needles of Stone*, I've had the details confirmed independently by another member of the group involved). This interesting piece of magical 'lock-picking' wasn't an exercise in altering people's *perception* of reality – as is the case with many magical operations – but of altering, locally, the *physical* definition of reality.

In effect, what are often called the 'laws of physics' are like any other 'laws' within English law: they describe guidelines to be interpreted, not final facts.

In effect, they describe what happens by default: the 'objective' world of public science is best described as a *default reality*. If we do nothing else, that is the physical reality we will perceive; but we *can* change our perception of it, and thus reality as we – and others – perceive it.

Another example: the Geller affair. Uri Geller was a much-publicised 'magician' (choose your own interpretation here) who demonstrated, under televised but suitably chaotic conditions, the 'magical' bending of metal objects such as spoons and watches. Whether he was faking it or not is actually irrelevant here: what matters is that he was convincing enough to be thought credible by respected authorities, who conveyed that impression to the people watching the programme. The result was that a large number of people found, overnight, that they could bend metal objects.

I was working with the British Society of Dowsers at the time and, despite the lack of any obvious link with our own work, many people came to talk to us about it. I remember, for example, one distinctly worried student who described how a spoon had wilted in his hand while he'd watched the programme; and a mother who said that her son had done the same, and had been able to repeat it at school the following day. She also said that he was bullied at school for years as a direct result: people, including children, do not like their tight definitions of reality shattered in quite such a visible way.

After a time, of course, the whole affair fizzled out: the 'default reality' re-asserted itself. But during the affair, for many people, this non-normal metal-bending was included in their definition of reality. For them, and for those around them, it was real.

We don't have to look far for other well-documented examples of the bending of the rules that make up the default reality: fire-walking, psychokinesis and poltergeist, to name only three. In most cases we do know some of the circumstances in which these phenomena are created: Batcheldor and Brookes-Smith's work on the creation of conditions for levitation of objects under full instrument recording, for example, was published by the Society for Psychical Research over a decade ago.

We don't know how these things work, and we probably never will; but we do know, to a surprising degree, how they can be worked. And that – how things can be worked – is all that a technology ever needs to know.

So if we are to create a magical technology, in which magic and technology are combined, we have these other tools at our disposal: we can change that 'default reality' if we need to. Which, if abused, as so many of our current tools are, could be a frightening prospect...

Changing the rules
Analytical thinking is what is still taught in schools as 'thinking'. It follows rules, precisely, to arrive at the correct answer to be deduced from those rules and the information it has to hand. In fact, computers do it better and faster than people can; so people have become worried by the implication that computers can think better than we can.

But this 'thinking' is hardly thought. An analytical program on a computer can *only* follow rules: it cannot create them. And a set of rules is simply a point of view, which itself selects the information which selects the point of view which selects the information... We have to be able to move outside of that cycle to see whether the rules we are applying are the best we could use.

I have a favourite dowsing example of this. One Australian author on the subject, who shall remain nameless, wrote a book describing what he called 'radial detection: a guide to the use of the radial detector, miscalled the divining rod'. 'Radial detection' was, he said, the sensation of specific physical radiations that he hadn't yet identified; and they could only be sensed by people who had no 'personal disadvantages': no scars, no fillings in the teeth, no glasses. (I've yet to find such a perfect specimen of middle-aged humanity: presumably he was so perfect).

He later discovered map-dowsing, which caused him some intellectual difficulties: he couldn't understand how these physical radiations from distant places could emanate from maps. Yet they must do, he said; but you could only map-dowse if you had no personal disadvantages (again), had the map and yourself aligned precisely north-south, beneath a bare electric light-bulb, at midnight, and you must have no clothes on... Apart from being cold, it's not a very efficient method of map-dowsing: not surprisingly, I don't know any map-dowsers using this method.

He claimed, though, that this was not so much a method, as the *only* way it could be done. We can see how his logic, and the limits of his point of view, built up a set of rules that prevented him from seeing any other way of doing it. And if we are to take a technological view of magic (or is it a magical view of technology?) we need to look at the ways in which we operate on the world, to build a set of rules for each circumstance, each occasion: a set of rules that is efficient, reliable, elegant (if you like) and appropriate.

We need to use ideas, beliefs, techniques *as tools*. We need, as a friend once put it, to build a 'cosmic toolkit' of ideas and rules and points of view to be used – and to be put back in the toolkit again once we've used them.

To do that, we need to know how to change the rules.

Changing the rules means a temporary end to analytic thinking. That kind of thinking can only follow one set of tracks at a time: if you need to change tools, you'll need to change your trains of thought. If the new skill is a 'magical' one, you need to review your assumptions about the default reality; and also to overcome the psychological barriers that Batcheldor labelled as 'witness inhibition' and 'ownership resistance' or, more simply, 'this can't be happening' and 'it isn't me doing this'.

The processes for creating and allowing these changes have been the main subject of magical and mystical study throughout the ages: this, if you like, is the real technology of mind. It comes up in any real study of science, too: Beveridge's chapter headings in *The Art of Scientific Investigation* include 'Chance', 'Imagination', 'Intuition', 'Reason: limitations and hazards'.

Scientific research has been described as 'one percent inspiration and ninety-nine percent perspiration': analysis and reason may provide ninety-nine percent of science, and all of its public image, but without that one percent there would be no science at all. One of the best-known examples in the history of science (also quoted by Beveridge) is Kekulé's description of his discovery of the ring-like structure of benzene – now one of the cornerstones of organic chemistry – at a time when all molecules were assumed to have chain-like structures:

> But it did not go well; my spirit was with other things. I turned the chair to the fireplace and sank into a half sleep. The atoms flitted before my eyes. Long rows, variously, more closely, united; all in movement wriggling and turning like snakes.
> And see, what was that? One of the snakes seized its own tail and the image whirled scornfully before my eyes. As though from a flash of lightning I awoke; I occupied the rest of the night in working out the consequences of the hypothesis... Let us learn to dream, gentlemen.

As any student of early chemistry will know, this choice of symbol was peculiarly apposite: Ourobouros, the serpent eating its own tail, is one of the key images of alchemy, symbolising an endless process of death and rebirth. Again, it's just a coincidence, like everything else: yet definitely one with a magical air to it.

There are, as you would expect, any number of methods for getting beyond the limitations of analytic thought. We can, though, break them down into three main classifications.

The first class set out *to knock the analytic mode out of action*, to force the train off the rails. Another, if imprecise term, is 'brainwashing': the analytic mode is hammered through any number of confusions into giving up its control. In political circles, the usual target is to put the mind 'back onto the right track', regardless of any damage to the train or its passengers. In magical circles, though, and particularly in mystical circles, the process is done with far more care, since there is no 'right' target to aim for. In these circles, 'alternative states of consciousness' may be induced through violent dance, through fasting, through drugs or, more gently, through insanely repetitive chants and movements: but always under the direction of a 'master', or at least of someone who has some idea of what is going on.

It works, but I feel it's rarely appropriate in the more practical realms of technology. Putting it bluntly, it carries more dangers than advantages: it's not worth the trouble and the risk, let alone the discomfort. And it seems a sad fact, in my own experience, that none of the groups I've met that worked with these techniques really had a clue what they were doing. The dangers are real: if you remain stuck 'off the rails', you're stuck in insanity.

The second class take a more gentle approach: they aim *to lull the analytic thinking to sleep*. They put the imagination to work: they suggest (rather than demand) parallels, comparisons, lines of thought, through analogy, allegory, myth and, perhaps most of all, through humour. Both the mystical and the magical traditions abound with a wry sense of humour, to suggest gently that there's always another way of looking at things, to show the limitations of 'normal' ways of thinking:

> Nasrudin was carrying home some liver which he had just bought. In the other hand he had a recipe for liver pie which a friend had given him.
> Suddenly a buzzard swooped down and carried off the liver.
> 'You fool!' shouted Nasrudin, 'you may have taken the meat – but *I* still have the recipe!'
> Idries Shah, *The Exploits of the Incomparable Mulla Nasrudin*

The same tools of analogy and imagery are used in technology, particularly in 'brainstorming' sessions and the like. For example, the current design of zip-fasteners for space-suits was derived by analogy from the Daedalus myth of threading a shell by tying a thread to the leg of a spider; and the peristaltic pump – a pump for corrosive fluids, with no moving parts in contact with the fluid – is based on the workings of our own digestive systems.

Another approach is that of 'twisted logic', in which premises and assumptions are moved around an apparently reasonable way, like pieces in a chess game. Each step will have its apparent logic: but at the end of each sequence of moves a very different game is in play. (In case you hadn't noticed, it's the main principle used in this study...)

And there are plenty of other tools, of course: as we've seen before, anything goes.

Meditation and similar techniques form the last class, which aim *to raise another mode of thought to dominance*. These are the hardest of all, but in many ways the most reliable: a slow process of meditation and the like can lead to a state in which the true/false judgements of logic become irrelevant. In the mystical tradition at least, this state is loosely described as 'enlightenment'.

To get to that stage, though, involves a total commitment and, usually, a total separation from the world in one sense or another, climbing higher and higher on some religious pole. And since, as technologists, as magicians, we are concerned with *working* on the world rather than viewing it in isolation from above, these approaches are probably not that important to us. Even so, the general techniques of meditation, for example, as tools for focussing the mind on one task, can be useful: as Leo put it in *SSOTBME*, 'all those boring books on meditation are roughly the equivalent of a chemistry primer' if the subject of study is the practical workings of the mind.

Most magical work plays games with the imagination, to realise and make real those images in the world that we – and others – experience. In technology, the imagination seems limited; but perhaps only because, as so-objective 'applied scientists', technologists have preferred not to show their human side. Yet that is insanity: for our technology *is* ourselves.

We should remember, too, that whatever we imagine is real – in imagination. It stays there until we realise it, until we make it real. We cannot cut bread with an imaginary knife; we cannot eat our words. (You can, said one magician friend, but they're not exactly fulfilling...) It's all too easy to get stuck in an imaginary world: the all-too-real world of nightmares, of paranoia, or of a dreamland that refuses to come into reality; or the frustration that comes when your skills fail to match your ambitions. And that, too, can lead to insanity – permanently.

So learning new skills, particularly those from the worlds of our imagination, we tread a delicate balance. We *know*, moving around the swamp of points of view, that we can go anywhere: but we have to do it with care.

This process really is 'the subtle art of insanity'; yet it leads us on to the practical art of magic.

The practical art of magic

Aleister Crowley, the magician and self-styled 'Great Beast', once described magic as 'the art and science of causing change in conformity with will': and most people would say that that is an adequate description of technology, too. There are practical differences, of course, between traditional magic and technology, but no real structural difference at all.

Above all, magical operations are just that: operations on the world as perceived. They are concerned with *doing* something: not just talking about it, or believing in it, but doing it. Which may come as something of a surprise if all you've come across so far is the concept that belief in magic is an aberration of the mind. And in a way, it is: as we've seen, any change to a point of view, wandering around in imagination, involves a certain degree of 'insanity'.

But our interest, in this study, is in merging magic and technology – putting them both into practice. There are two sides to this: technology as 'applied science' has lost its magic, but gained in clarity; magic, as a set of technologies beyond science, has become isolated in a world of its own, and has lost its sense of purpose in an everyday world, its sense even of being technology. Each side has its own traditions, its strengths and weaknesses, with much to offer to the other; and much to be seen beyond.

A sense of vision

The first skill we have to learn is how to see. Not just with the eyes: but with vision, with the 'mind's eye', like Kekulé's image of molecules as snakes, twisting and turning in the fire. Creating images – using our imagination – and letting those images return with more detail than we gave them at the start, we can learn to navigate our way through the swamp of ideas, and pick out those points of view which might be useful.

Let's start, then, with one particular image, from another of Idries Shah's 'Nasrudin' tales:

> Someone saw Nasrudin searching for something on the ground.
> 'What have you lost, Mulla?' he asked. 'My key', said the Mulla. So they both went down on their knees and looked for it.
> After a time the other man asked: 'Where exactly did you drop it?'
> 'In my own house.'
> 'Then why are you looking here?'
> 'There is more light here than inside my own house.'
> Idries Shah, *The Exploits of the Incomparable Mulla Nasrudin*

If you're looking for ideas, you're not likely to find them where they are not: 'discoveries are beyond the reach of reason'. To find ideas where they are, you need to look where they are; and, going back to the swamp analogy, that means that we need to get away from the platform for a while, for we'll find nothing new there.

So let's develop that image a little further. The description of science, of 'reason', as a solid platform is a useful one, because it implies safety – 'safety first' is a good rule in any technology, including those from traditional magic, so it's a good idea to have a known, safe refuge when things start getting a little insane. Once we know the platform, our safe refuge, we can move from it, and return to it in imagination at any time. So start by building your own image of this platform, this safe refuge:

In our earlier analogy, the scientific world-view was likened to
a platform of inter-related points of view within the
swamp. Imagine that you're on this platform, a solid, safe
and fortress-like structure of impenetrable logic and
unassailable rightness.

Look at the walls of reason around you: they defend you
from the unknown world of the swamp outside.

Look at the walls around you: they also enclose you,
preventing you from seeing the world outside from other
points of view.

Let images arise – or not – as they like: let this world
describe itself to you.

You want new ideas from outside of the fortress – there's
more light out *there*.

Imagine that you can create a window in the fortress wall in
front of you. As you imagine it, looking at the wall, the
window appears. Just let this image arise by itself.

See what kind of window you have created, looking out over
the swamp. This is *your* window, into *your* world.

Sunlight comes through the window; a butterfly flutters past.
Through the now-open window comes a scent of wild
flowers; a different world.

Come closer to the window, and look out. Describe to
yourself your vision of other points of view, as you can
see them out through the other side of the window.

Now close the window, and return.

This is a game called 'imaging', a classic tool of traditional magic now much in favour with the self-development schools. Like any game in technology, it is a tool with a purpose, leading somewhere – but only you will know where.

It's a useful skill to develop: a trained and directed imagination. If you're told that something you want to develop has been proved to be impossible, remember de Bono's phrase: 'proof is often no more than a lack of imagination'.

There's a knack to it, as with riding a bicycle or any other practical skill; so it may take a little time before you can move around easily in this world of your imagination. The point of this game is to put your imagination to *use*: daydreams with a purpose. Until you get the hang of it, you may find it easier to get someone to read these images to you, quietly and carefully, so that you can follow them in your mind's eye without the visual clutter from reading it yourself.

> Imagine that you're on the fortress-platform of the scientific world again.
> Return to where you created the window; notice any details that have changed since you were here last.
> If the window is no longer in the wall, re-create it. Again, sunlight comes through the window; a butterfly flutters past; the scent of wild flowers comes through.
> Note if anything seems to have changed; note if anything seems to be informing you of its presence.
> Since all is well, open the window wide.
> Look right out of the window; lean right out and look from side to side. Note in your mind what you can now see.
> Move back inside: close the window.
> Now change the window into a glass door, opening out onto ground level beyond the fortress wall.
> You can open this door at any time. If you want to, open the door and look out again: though don't leave the fortress yet.
> When you've finished, close the door, and return.

In imagination, you now have a door into a different world. In traditional magic, people approach this world as a 'quest': a place in which questions can be answered, and answers can be questioned. Before we go there again, build up in your mind, carefully, the details of something that you've been trying to think through – anything will do – and that you'd like to see from a different point of view.

Return to that glass door, in the fortress wall of the platform of science.

Quietly, and with confidence, open the door.

Before you go out, you see that there is a coil of rope beside the door, tied to a ring: a safety-line. With this tied to your belt, you know that you can return, on your own, at any time.

Tie the rope to your belt, and move out.

Take some time first to look at the walls from the outside, from a different point of view: look at any places on the outside of the wall that are in need of repair.

Then look slowly around you. You won't need to create the world outside, but it may need some help from you to give it shape and form. To help it, keep in mind the quest with which you came here.

Spend some time looking around at the edges of the platform, outside of the walls.

Look out into the distance, and see the old buildings of some abandoned parts of the platform, like astrology and alchemy and herbal medicine; they're in better shape than you thought, and there are people living and working there.

You can see people around, in the distance, but you do not meet them.

You are safe: you have your safety-line back to the fortress.

You came here with a quest in mind: take some time to note if anything about this quest comes into your view.

When you have seen enough, return back to the fortress.

Close the door; untie your safety-line; and return.

It's worth while writing down anything that comes to mind when wandering around in this imaginary world. If someone is reading these images to you, tell them what you see. It does not need to *mean* anything at present – although it's surprising how often the images arising from this game are immediately useful.

A couple of other points are worth keeping in mind. The first is that the door, although it may be on the same place on the *inside* of the fortress, may well open out onto a different place *outside* each time you go through – to enable you to look at things from a different point of view. The other is that you should not just rely on visual images: note any sounds, smells, sensations, the general 'taste' of what is going on. Although you create, by describing it to yourself, the loose structure of this world, your imagination goes on to fill in the details: and it's in the details that you'll find the information that you need for your quest, and that you can use when you return.

So before we move on, you might like to go back to the previous image, and look through it again.

Introducing the guide
So far, we've played safe. We haven't moved without a safety-line – just like the technologists in the swamp analogy. It's time we spread our weight a little, and moved out onto the swamp itself.

Again, we still need to play safe: so we'll ask for a guide. If you've read any classic magical literature, you'll no doubt recognise this part of the game. But to take a slightly different line from the classical one, we will declare that *this guide is an aspect of ourselves*: it is *not* something separate. If you like, it is an aspect of you that knows how to move freely around in the morass of different ideas and images – so it knows how to guide you, safely, from one view-point to another, to look at them and from them as you need.

Return to the glass door.
It is already open, waiting for you.
In the doorway stands your guide, waiting patiently. Note what form it takes; its form may change from time to time, but it will always be an aspect of *you*, to take you to where you need to go.

At first with caution, your guide leads you out over the swamp. Take some time to note how you move, what images you move through.

You make a short stop to rest at a well-restored building close by to the fortress: the world-view of astrology. Here things are seen not as cause and effect, but as events that parallel each other, different results of greater causes: 'as above, so below'. Look at the world for a while from this point of view.

Your path leads on to a tightly enclosed maze: the logic of the computer program we looked at earlier. You can watch as each program instruction passes by: note the circumstances under which each instruction occurs, look at things from the program's point of view.

It follows each instruction methodically, relentlessly, following the fixed paths of its logic; you watch it a while, and move on.

Your guide leads you now to show you another place, another point of view, which has some bearing on the quest with which you entered this world. This place is *your* choice; what this place means to you depends on *you*. You can spend some time here questioning the answers – whatever form they may take – which the place can show you.

You now return with your guide to the fortress, to the glass door.
Take your leave of the guide; close the door; and return.

We do need to be clear about the nature of this guide: it is *very* important to view it as being an aspect of ourselves. In spiritualist circles, where the same game is played in a rather different guise, the 'spirit-guide' is always seen as an entity separate from the participants in the game: yet while this does in some ways solve the problem of 'ownership resistance', I regard this approach as dangerous in the extreme, since it abdicates all control and responsibility to an unknown and undirectable but very real force.

This is not a trivial point. In the 'Philip' experiments in Toronto, a university research team invented a non-existent character named Philip, complete with an imaginary life-history, and then 'called up his spirit' in séance-style experiments. Despite its imaginary beginnings, the entity became real enough to respond with standard séance-type phenomena: rappings, table-tipping and even minor poltergeist events – this was even demonstrated live on television. Slowly, though, it began to have 'a life of its own', and in the end was sent on to an imaginary next world with something akin to an exorcism ceremony. All this from a character that had never existed in the 'real' world at all.

Since we are working in an imaginary world in this game, it's as well to remember that *imaginary entities are real in that world* – and, without some semblance of control, can, on occasion, be dangerous. Your guide is your choice, and your choice is also your guide: as the phrase goes, 'do what you will – but be very sure that you will it'.

A world of archetypes
Another way of looking at these 'entities' we meet is as 'archetypes': collections of aspects of ourselves, or even of places or objects or concepts, making up a describable 'personality'. Having personality, they have a kind of intelligence: you can ask them questions, and they can respond in their own way. You can even put yourself in the position of that archetype; if you like, 'be' that archetype; experience things from that point of view.

Beveridge cites several examples where researchers have imagined that which they are studying as having a personality, so that they could look at things from its point of view; and we've already seen a similar process used in debugging a computer program. In the magical field the process is referred to as 'invocation' – literally, calling within – calling up this specific group of attributes, of characteristics, in order to experience them. An example of the use of this is the skilled actor, who wears a part, a character, in the same way that we ordinary folk will wear clothing. The trick, having taken on the character, is to be able to put it down again... And that's where we rely on our imaginary guide for help.

These archetypes can be anything at all, and can be allocated any sort of name: a name that is simply a working label for a set of characteristics. In effect, this is the same as the computing principle of 'if you don't know what it is, give it a name'. An eccentric magician friend, for instance, once invented 'a pair of very sticky goddesses', Uhu and Araldite (which happen to be the names of two commercial glues) for invocation by his students. It's more practical, though, to pick on a known name or label, and build on that.

While the best-known labels come from the magical tradition, with its pantheon of gods and goddesses, angels, demons, devas and the rest of the magical menagerie, we can choose labels from *any* tradition or field of study: we create this imaginary world, so the labels are ours to choose. Anything goes. So we'll select two to enquire of in our next excursion into the swamp: the Hanged Man from the Tarot cards, a classic symbol with a not-so-obvious interpretation; and that shadowy concept from present-day physics, the Quantum, the minimum 'packet' of energy.

Return to meet your guide at the glass door.

Together, you move out over the swamp. If you can, keep track of your path through this maze of images and ideas.

After a short time, you arrive at a clearing in the swamp.

Here you see an upright wooden frame. A young man is suspended upside-down from the frame, with his head just clear of the ground. He is certainly alive and well; relaxed, even. One ankle is tied to the top of the frame, and the other leg crossed over behind the knee. His arms are behind his back: we cannot see whether they have been tied there, or if he is simply holding them in that position. Despite this peculiar predicament, his face has an expression of calm detachment, a kind of distant awareness; you can sense a glow, almost a halo, around his head.

Move closer. Take in the details: the shape of the frame, the clothes he is wearing, and so on.

Move closer still: move right inside this image. See things for a while from this strange upside-down point of view.

Now return to be beside your guide, looking again at this figure from the outside.

Move on to another place, a place of total but simultaneous contrasts: everything, it seems, is both black and white at the same time.

At first, you cannot focus on anything: everything is in movement, yet not in movement. It seems you can tell where something is, but not where it's going; or where it's going, but not where it is. Always, but simultaneously, one *or* the other.

Each movement or not-movement is a single packet of energy, a quantum of energy. As an object, a 'particle', it is static; at the same time, as energy, as a 'wave', it is movement, and only movement. For a while, be with this flow and not-flow: resolve this impossibility of opposites by *being* the quantum.

Now return to yourself and your guide.

Together, you move on, back to the fortress.

Take your leave of the guide; close the door; and return.

You've probably noticed that we've made a definite point each time of closing the door: almost a ritual. Part of it is plain tidiness, and an element of safety as well – you don't want some imaginary demon from your imaginary world wandering around loose in the fortress, after all – and part of it *is* ritual, to reinforce the sequence, to make the whole process automatic.

Another way to see ritual is as a pre-planned sequence of events, as a sequence with a purpose, everything just *so*. Rituals are an essential part of any skill, in some form or another: the pre-take-off checks in an aircraft, for example, become almost an automatic part of any pilot. The responses and reactions we develop when we learn to drive a car become quite unconscious, quite automatic – and are faster as a result. You pick them up, select them, as you get into the driving seat, almost as a matter of ritual; and you put them down again when you leave.

Being automatic, but with awareness above, you see immediately any break in the sequence. The trick is to avoid being stuck in the ritual, doing it for its own sake; but rather to learn to see its *use*.

The main function of rituals in the magical tradition is to set up some particular state of emotion or awareness: magic, as we have seen, can be described as a technology of the mind. Like the checklists of the pilot, or the pre-operation wash-and-dress of the surgeon, a ritual of any kind sets up specific conditions – a specific context, if you like – in both the operator and the 'real' world as they intend to perceive it. Its other function, more prominent in the conventional technologies, is to highlight any break in the sequence, anything which doesn't fit: so you can see and act on this changed context accordingly.

So a ritual is a functional tool, like any other: it may not seem much like a screwdriver or a spanner, of course, but the intention behind it is much the same.

Getting unstuck

Magic and technology are above all practical arts; so this game that we've been following does indeed have practical use. Its function is to provide a solution to 'stuckness', where you can't think of a solution to a problem: the world has given you an answer, so what on earth was the question that that answer replied to? In that kind of situation you cannot be reasonable: the answer is outside of the neat web of logic that you've so far built. You're floundering around, trying to think of *the* fact, *the* bit of information you've missed, that would give you the clue you need. As Pirsig put it:

> I keep wanting to go back to that analogy of fishing for facts. I can just see somebody asking with great frustration, "Yes, but *which* facts do you fish for? There's got to be more to it than *that.*"
> But the answer is that if you know which facts you're fishing for you're no longer fishing. You've caught them.
>
> Robert M Pirsig, *Zen and the Art of Motorcycle Maintenance*

And that applies to *any* skill; indeed, we could say that the ability to find out new facts about the work and yourself *and put them to use* is a practical definition of skill.

When you're stuck, you need some inspired guesswork. And that's what our game is: practical inspiration. If you're stuck, describe to yourself, in detail, what you think the problem is. Then go to the fortress, go through the door, and ask your guide to take you to meet someone or something that can show you some new light on the problem.

Do it. Oh, and don't forget to close the door when you return...

Inspired guesswork

Let's look at this a different way. If you're faced with the problem, as maintenance men are, of knowing that there's a leak *somewhere* in a mile of buried pipe, where would you start? The obvious way – the scientific way, if you like – involves digging up the whole mile of pipe: not exactly the quickest way of tackling the problem, but certainly thorough. Before you spend a large amount of time and effort on the old 'brute force and ignorance' approach, you could try a little inspired guesswork.

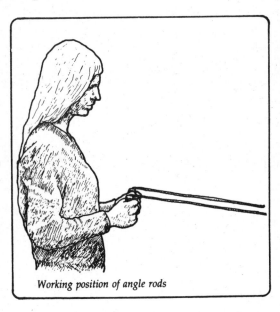

Working position of angle rods

First, make a pair of dowsing rods out of a couple of old coat-hangers, and hold them as shown in the illustration. They must be able to move freely; the trick here is to hold them just below the horizontal so that they can balance easily, yet move a lot if your wrists rotate a little.

Now we play a game similar, in its way, to a computer program. You declare – verbally, if you like, but more usually just to yourself – that the rods will cross over when where you are (marked by your toes) coincides with what you're looking for. Build up in your mind an image that you're looking for is the pipe: describe it to yourself in the same way that you looked at that orange earlier; describe it to yourself in the same way that you framed your 'quest' when visiting the imaginary world beyond the fortress. And that image now matches what you're looking for: so do it.

Play with it. Remember about 'beginner's luck': the real trick is to get there, and stay in that state.

Finding the pipe is one thing: what about the leak? One way is to add another instruction to the 'program': that the rods will open out, rather than cross over, when you're above the leak. The skill here is to learn to keep yourself open: as in the world beyond the fortress, build an image, but note how the image can itself add details for you to see and use. As Pirsig's analogy puts it, you're 'fishing for facts' about the world: work on the world, and listen to what the world tells you in return.

And if it 'didn't work'? Well, try again; or rather, play at it again. If it still doesn't want to play, there's always the 'brute force and ignorance' method – you haven't actually lost anything by trying it out.

The tool for the job
One of the points we do need to learn about these tools from the magical tradition – or any tools, for that matter – is to use them in an appropriate way. If you're working on a world that you know, it's usually easier to use a way that you know. It's only when you're getting stuck, or the method, the tool, that you're using begins to be inefficient, unreliable, inelegant, inappropriate, that you need to look elsewhere. And you need to keep your awareness wide, to let the world tell *you* that what you're doing, the way that you're working, is indeed inefficient and the rest.

Dowsing, for example, has limitations: it depends perhaps too much on your own skill to be regarded as a reliable technology, a reliable tool. For most detection work, you're better off using some ordinary physical means of detection: a metal detector, for example. But metal detectors become unreliable if what you're looking for is more than a few feet below the surface; they can only find specific materials (and not the plastics used for most water-piping of these days); and they can go wrong in quite surprising ways.

My favourite example of this was an incident during the building of the Humber suspension bridge. Careless parking of some pre-cast concrete bridge sections in the store area crushed a main water supply pipe. Unfortunately, this was the supply to the Hull fish market, which had to be closed down immediately. The break was, of course, *somewhere* in the several acres of the storage park: no real indication as to where.

The plumbers from the docks arrived with their toolbags and their fencing-wire dowsing rods at the same time as a van-full of electronic search equipment from the Water Board. The technicians looked at the plumbers' equipment with disdain: very crude, they thought. But when they turned on their sensors, all they could pick up was Radio Humberside – which was right next to the field. By the time the technicians had disentangled this little tuning problem, the plumbers from the docks had not only found the leak, but mended it...

On the other hand, I've often found dowsers using their skills in quite inappropriate ways, looking for things by dowsing when a physical tool – or a little common sense – would do the job better. One example that always annoys me is when people ask me to wave a dowsing rod over a glass of water to 'prove' that the rod will react when over water. But I already know that it's water: I can *see* that, without needing any fancy apparatus – so why bother?

The most likely end-result of parlour-games like these, games without a purpose, is that they don't work, they don't prove anything – because there's nothing to prove. 'Proof' is something that science can concern itself about, while we are only concerned with technology, the use of tools: and the only value of a tool – any tool – is in its *use*.

Building a toolkit

If we view ideas and techniques *as tools* rather than as 'facts' to be proved true or false in an arbitrary logic, we can go back through the literature of the magical and mystical traditions again with a very different intent: *to put them to use*. In technology, in this magical technology, *anything goes*: anything is 'true' if it can be put to use in practice in our perception of the world.

See astrology, for example, not as a collection of facts or fictions, but as ideas and conceptual tools to be put to use. On some occasions, they are useful; certainly astrology's key concept of 'as above, so below', of parallel events without causes, is useful to help us move away from our current culture's obsession with a spurious concept of cause and effect.

Its images and archetypes are some of the most descriptive around; and in a good practitioner's hands, modern astrology can be a remarkably accurate tool for describing the processes of a person's constant transitions from one state of being to another. As a tool to describe *events*, it has always been dubious at best; but as a tool to describe the *context* behind events, that's quite a different matter. Knowing its strengths and weaknesses, we can use it: but in a way that should be efficient, reliable, elegant (if you like) and appropriate.

The same goes for other tools, other games. An amazing amount of work has been done, and then forgotten; researched, written down, checked and re-checked, archived – and yet never put to use, simply because it didn't fit with the rest of our views about what technology is and does.

For example, we do not know how psychokinesis works, and probably never will: but we do know enough about *how it can be worked* – how to create and apply the energies involved – to create ways in which we can put it to use. From Batcheldor and Brookes-Smith's work, the principles are stunningly simple: use a little dim light (to get over 'witness inhibition') and a small group (to get over 'ownership resistance'); use a 'joker' (a member of the group who fakes things up a little) to give some encouragement; then go off to change the default reality for a while. The practice, of course, is a little more difficult: reality is never quite that simple.

Another example: alchemy. This has always concerned itself with transmutation of elements in both inner and outer worlds: so it could, perhaps, be applied to give us a solution to the problems of nuclear waste bequeathed us by a short-sighted view of technology as 'applied science'. Levitation too, perhaps: there is a surprising amount of sense in Douglas Adams' wry description that 'the way to fly is to throw yourself at the ground – and miss'.

With an awareness of magic, we can put chance and coincidence to practical use. And we don't need to worry about trying to invent causes for events, to make them 'fit' reality – because we've invented that reality in the first place. We all know those magical occasions when everything was 'meant' to work, when all the right people turned up at exactly the right time: now we can put that sense of magic to *use*, as part of our more magical approach to technology.

Again, we can direct the process, the chances, the coincidences: but we cannot control them. If we try to, the magic just vanishes. This seems always to happen if one person in a team claims that sense of power for their own, or claims to have a monopoly on 'the truth'. As one friend put it, 'a quick trip up to the astral to grab a slice of the action' tends not to get very far: that sense of magic is a little more elusive. As with any technology, we need to be aware of whether what we do is efficient, reliable, elegant (if you like) and, above all, appropriate.

If 'anything goes', it's up to us how we do it, up to *us* to decide what is an appropriate tool for the job. Using the technology of mind that we can find in the magical tradition, we can turn that paradox of 'things have not only to be seen to be believed, but also to be believed to be seen' on its head: by moving, with care, our definition of that 'default reality', we can, it seems, create any world that we want.

But what kind of world do we want? Perhaps more to the point, what kind of world do we *need?*

Before we can create any new world, before we can make real some imaginary utopia, we need enough vision to see what really *is* needed; we certainly need more vision than the blinkered arrogance of our present 'applied science'. We need, if you like, a better awareness of the miraculous in the ordinary, to bring the magic of life into our technology; and to reduce our arrogance, of 'God made in the image of Man', we need, perhaps to be a little more humble in our approach to nature and its realities.

And, in the magical tradition, there's no stronger image to show us the foolishness of our current 'wisdom' than the Tarot's symbol of the 'wise fool': the joker in the pack.

The joker in the pack

The joker, the jester, the 'fool who goes where angels fear to tread', is a symbol that permeates the whole of the magical tradition. And it's a symbol that any technology would do well to adopt and understand.

There are many different aspects to the symbol, all of which are valid to us in learning to work on the world as it is. One is the sheer stupidity of following a path without looking where it takes us – as is the case with the Tarot fool, who is usually shown about to walk straight over the edge of a cliff.

Another image is that of the Mediaeval 'Lord of Misrule', the anarchic symbol of change – which can be directed, perhaps, but certainly not controlled in our accustomed scientific manner.

Humour, too, leads us gently to find new ways of looking at the world; the court jester is able to show us things, through humour, that we would perhaps rather not see: our vanity and our pride when we blunder our way into some new skill, for example.

And nature itself plays at the joker, tricking us when we least expect it: without an awareness of nature and our own nature, life can become distinctly unamusing.

'Applied science' – a technology without magic

One of the saddest disasters that can befall a technology is for it to be treated as a science. There is plenty we can learn from methodology, the study of methods – but only when they are studies *in practice*. As soon as we think in terms of 'theory and practice' – theory *before* practice – we can be sure that we are going to hit trouble as soon as we meet up with anything that doesn't fit the assumptions of theory: and we've already seen that we can never, in practice, analyse enough factors for theory to be able to predict *anything* with total reliability.

Theory can only tell us where we've been, not where we are or where we're going: it's a rear-view mirror, taking us nowhere but backwards on some neat train of thought. It's a very blinkered view of where things have been: it can only see what it expects – or rather expected – to see. And by assuming that what it sees is 'objective truth', it leaves no place for subjective experience, for the other multitudinous factors that make up the context of the real world we experience.

This view is bad enough in fields where the human factors are less crucial, or at least less visible: in engineering, for example, where a design (the theory, if you like) which fails to allow for erratic workmanship and variable materials can be fantastically expensive in the long run – the now-ageing first- and second-generation nuclear reactors now littering the country being a good case in point.

But it's in the so-called 'soft sciences' or, more accurately, the human technologies, like psychology, sociology, ecology, economics and the rest, that this attitude becomes downright dangerous. If the technology only knows about 'truth', with no concept of value, of quality, then it has no room for a sense of the quality of life.

A technology that regards scientific truth as its sole criterion can only understand value in terms of quantity – 'bigger is better'. To that particular point of view – and it *is* a point of view, as we have seen, and not 'fact' – any other sense of value is, by definition, 'subjective', and therefore unscientific.

So it's not surprising that the other, more qualitative view – 'small is beautiful' – is seen by some as magical, and by others as plain unscientific. Which, of course, it is: both magical and unscientific.

But that – complaining that magic and technology are not scientific – is precisely the mistake that concerns us. They never were science: it's true that they *use* scientific concepts in practice, and they *use* scientific-style logic and reasoning in their methodologies, their study of techniques, but that does not bind them to see things only from that platform, that group of points of view. They put science to use, so in that sense they are 'applied science'; but they put *any* idea, *any* concept to use, as long as, and only for as long as, it is useful.

There are two aspects of science which are definitely *not* useful. One is the limiting effect of its constant rearward thinking, which can make it anything from difficult to impossible to apply new ideas and new approaches and put them to use. The other, more serious, aspect is a side-effect of its true-or-false logic: if something works, it is true, therefore, logically, any other method is false, wrong, forbidden. This method works, *therefore* there is no other way it can be done; *there is no alternative*, it says.

When the true-or-false concerns of logic have priority over value – as they do in science – what we get in practice is not technology, but something more like a mess of dogma. As another Nasrudin tale puts it:

> Nasrudin was throwing handfuls of crumbs around his house.
> 'What are you doing?' someone asked him.
> 'Keeping the tigers away.'
> 'But there are no tigers in these parts.'
> 'That's right. Effective, isn't it?'
> Idries Shah, *The Exploits of the Incomparable Mulla Nasrudin*

Once a technology becomes 'applied science', it seems, people forget to *think*: about the wider context, or about what they are doing in practice. It also seems that they forget how to laugh, how to use humour as a means of moving from one point of view to another. Our technologies, our ways of working on the world, are a serious matter, it's true: but when the humour goes, when the joy goes, so does the humanity – and with it the magic of technology.

Medical madness

Perhaps the saddest examples of this can be found in the chaotic shambles that we call medicine. There are any number of ways of looking at health, any number of points of view, every one of which claims to be the *only* way of looking at health. You could say that we understand extremely well in medicine why people fall ill; what we don't really understand at all is why they don't.

The conventional medical approach regards disease in terms of cause and effect: known causes, such as bacteria or carcinogenic agents, produce known symptoms, known effects on the health of the patient, from which the cause is diagnosed. Symptoms are treated with drugs which, unfortunately, may have side-effects, other symptoms which have to be treated by yet more drugs: a cycle which seems to profit the doctors, the 'caring' professions, the drug companies, their advertising agencies – everyone, in fact, except the patient.

In effect, the problem with this approach is that causes are seen in terms of symptoms, which, by reverse reasoning, are assumed to be the result of the causes. Diagnosis, like scientific investigation, is an art, not a science: its 'discoveries are beyond the reach of reason'. Without awareness – something that logic alone cannot provide – things can go badly astray:

> Imagine that you're wearing a tight collar.
> Now, if I pull hard at the back of the collar, where do you feel the tension?
> That tension is a symptom of the disease: so I'll treat it by pulling hard on the collar at that point where you feel the tension.
> Strange: the tension has gone from the first place, but it's started again somewhere else. So I'll treat you by pulling hard there as well.
> Now, you're feeling fine, aren't you? Aren't you...?

Above all, this kind of medicine believes that it is scientific, 'objective'. The patient is literally 'one who waits patiently', without responsibility for their own health: disease is objective, so treatment is 'done' to them, actions in which they are not involved other than as an obedient lump of wayward flesh. Any other approach to medicine is wrong, because this one works on *facts*.

But that is precisely the point: this approach cannot work simply because there are no facts – only experiences. To quote Beveridge again, "most biological 'facts' and theories are only true under certain conditions and our knowledge is so incomplete that at best we can only reason on probabilities and possibilities". This is obvious to anyone working in the real world of general medical practice, as my parents did: every patient was different, so the techniques used – although strictly speaking those of conventional medicine – were different for every patient.

Yet in the sterile atmosphere of hospital wards and laboratories, this view would no doubt be thought of as unscientific, as dubious medicine: an objective world-view has no place for the anarchic differences between individuals. Technical skill and expertise may be one of the hallmarks of conventional medicine; but vision, an awareness of what medicine *means* in practice to the patient, is often conspicuous only by its absence.

In principle, then, you would expect vision and awareness to typify 'alternative medicine'. And yes, after a fashion, it does: being points of view in what is actually a magical technology, acupuncture, homoeopathy, herbal medicine, dietary therapies and the rest certainly work for many people. And people are seen as involved *people* rather than passive 'patients': health for the whole being. Routinely, though, you'll find that the practitioners of these therapies argue with their colleagues, other practitioners, and doctors from conventional medicine, as to which therapy is more 'true', is *really* more scientific.

But in practice this is pointless: all of them are true, in the sense that all do work, after a fashion; therefore, logically, none of them are true in an absolute sense, since they all disagree as to how things *really* work. 'Everyone is always right, but no-one is ever right'. Being right doesn't necessarily help if it fails to see the whole of the problem: surgeons' reports in the nineteenth century routinely ended with the phrase "operation successful, but patient died", which seems a somewhat extreme example of missing the point of medicine in the first place.

To make our medicine work as a technology, we need to be able to select ideas from each approach, from each view of medicine, according to the practical needs of the patient: not, as so often happens at present, according to the dictates of theory. Above all, medicine needs to be *appropriate* to the individual: efficient, reliable, elegant and apt. To bring the sense of meaning back into medicine, it needs to be a technology with magic: a magical technology.

'Murphy's law'

Science describes many so-called laws of nature, that describe how the world really works; and as we've seen, they're not so much laws as guidelines that we can use to create reality – guidelines that describe rather than define. Yet to engineers, and to anyone working on the real world in a practical way, there *is* one real immutable law of nature. Often known as 'Murphy's law', it states, simply and baldly, that: 'If something can go wrong, it will'.

And that *is* a fact: not of theory, but of experience. There's always *something* that doesn't fit, some combination of factors that you didn't or couldn't allow for. There are innumerable variants on the theme: one of my favourites, from the computing field, states that 'As soon as you think you've made your program idiot-proof, along comes a better idiot'. Nature plays the joker, the jester who teaches through jokes that strike home where we would often rather that they did not: if we don't learn the lessons, we end up being the fool.

Which is probably why skilled workers tend to approach the world in a cautious way, with a wry sense of humour. Watch any craftsman at work in any real skill – be it carpentry or chemistry, dowsing or programming, or whatever else you choose – and you'll see much the same expression on their faces. At the point of action, there is total attention: concentration on the work in hand, yet also wide open, sensing the context for difficulties or dangers coming from any direction.

On our swamp analogy, they're at one specific point of view, yet spreading their weight, sensing when it's time to move on. Like an artist, or the stand-up comedian, it's all in the timing. So there's no place for the 'brute force and ignorance' school of non-craftsmanship: by failing to listen what is happening in the swamp – to what is happening to the world and the work in front of them – they miss their timing, miss their footing in the swamp. And it can take a lot of floundering in the mud to make a macho-style male realise that nature doesn't suffer fools gladly.

If we're lost in the swamp, we're stuck – we don't know which way to go. In debugging a program, we *know* the program has followed the logic, the points of view, that we gave it: so which way did we tell it to go? A scientific approach, with its closed logic, can only tell us the way back from one point of view: more often than not it will only lead us round in circles. Being stuck, trapped in a circular pattern of thought, is distinctly unfunny: but we can use humour to lighten the way out. Or images, or analogies, or any of the other tools we looked at in the previous chapter: anything which will let us see the world as it is, rather than as our point of view assumes it to be.

Mistaken magic
If 'applied science' is a dangerous disaster, attempting to treat the technologies of traditional magic in a scientific manner is even worse. If we accept that that magic is a technology of mind, it is obvious that to try to understand it from an 'objective' point of view, one that does not even recognise the involvement of the mind, will lead to fundamental problems: it simply isn't going to make any sense. Trying to use a true/false logic in an area of experience in which everything and nothing is true at the same time, and in which only value judgements are valid, is, bluntly, insane.

And yet people do it. We've already seen the problem with the 'alternative therapies' in medicine, in which some controls do at least exist; imagine, then, the dangers where no safety devices exist at all.

One example that comes to mind is 'channelling', a formalised version of the imaging we saw in the last chapter: it's a key part of both spiritualist-style and so-called 'New-Age' operations. A 'spirit guide' or 'guardian angel' (the terminology varies) is contacted – imagined – and provides information on how to handle some problem. So far, that's no different in principle from what we've done here, talking to our imaginary guide; but the interpretation is quite different. Instead of treating the information provided as just that – information – it is treated as fact: final, absolute Truth, from a 'higher being' or a 'higher consciousness'. With no way whatsoever of checking the validity of that 'truth'.

On occasions the information is useful – if you like, 'inspired' information in the same way that finding a pipe by dowsing is 'inspired guesswork'. Information is always useful if you can put it to use. But there are no control procedures, no 'noise filters': no way of separating out the useful information from the junk that comes with it, junk from any number of possible areas in the human psyche and way beyond.

In magical technologies, there *are* no facts, only information and its interpretation: yet in these cases everything is treated as fact, simply because it comes through with an emotional loading, an apparent authority – a jumble of information and non-fact which is in practice determined by whatever interpretation happens to be lying around. One result of this indiscretion is that in the Glastonbury area, where I have lived for some years, we seemed to have been besieged by an endless stream of self-appointed Messiahs, each of whom was convinced, with deadly seriousness and a distinct lack of humour, that they alone had the divine guidance to lead the people to grace...

For a while, it's almost amusing: a bitter cosmic joke. Listening to the sound of chaos, as the over-inflated ego of yet another fool is punctured, may provide us with a grim and all-too-regular entertainment; yet the sheer waste of mental resources, the destruction of minds that can take years to repair – that isn't amusing. Magic without the awareness, the discrimination, of technology is definitely no joke at all.

It's with very good reason that, in the Tarot, the Fool is the key card, the joker in the pack. A card without number, both beginning and end, it symbolises our progress through this maze of realities. An endless exploration: and each time we return to some place, we know it for the first time. With 'beginner's luck' and a beginner's joy of discovery, we can always learn anew: and there is always much to learn.

Nature has the last laugh

Within our technology, we have another lesson to learn: that we have to learn to live with nature and our nature. We are part of nature: we cannot control it without controlling ourselves – which we are manifestly unable to do. We can, perhaps, *direct* what goes on, though we cannot control it: the distinction is important.

Nature, the world we live in, is not an inanimate 'thing': seen as a whole, it is a living organism of which we are part. This concept has always been part of traditional magic, perhaps because of its pagan roots, perhaps also because of its greater concern with an overall awareness; it also comes through in more recent scientific research, such as James Lovelock's work on the Gaia Hypothesis – that the condition of the Earth is *actively* made fit for life by life itself, working interactively as a whole. An appropriate choice of name for a scientific hypothesis: Gaia is an ancient Greek name for Mother Earth.

It's interesting to follow through the old pagan concepts in terms of what they mean to a technology that is, as we have suggested, concerned with being efficient, reliable, elegant and appropriate: because if it is working with nature, in the real world, it has to be reliable over aeons of time – it has to be maintainable indefinitely. Which, it's clear, can hardly be said of our present technologies developed from the blunderings of a half-blind 'applied science'. We have seen that we cannot be scientific about our relationship with the Earth: there are too many factors involved, so in a sense we have to be unreasonable, to feel and to sense our way through to apt solutions.

And, historically, we have had what appears to be help from the Earth, from Gaia itself, in reaching towards sane solutions. From the work of that eccentric cataloguer Charles Fort in his *Book of the Damned*, and others such as John Michell and Bob Rickard with their listings of *Phenomena*, we can see that we've been sent all manner of mad happenings – from showers of frogs and fishes, to bizarre horrors like spontaneous human combustion – to show us that nature always has other points of view.

Throughout history, we've had muddled messages from an assortment of perceived yet imaginary entities – interpreted as fairies, angels or spacemen, depending on the point of view from which they're seen – that have, it seems, been visited on us by nature itself. The only common factor in these insane events (and there must have been thousands of them throughout history) is their sense of authority: the Angel at Lourdes, the 'man in black' who handed over the design for the Great Seal of the United States, and many others. Somehow, 'visitors from inner space' seems more likely than otherwise: but what 'inner space'?

In any case, 'there are more things in heaven and earth, Horatio, than are dreamed of in your philosophy': a painful truth to discover, as many of us have found out... The 'Old Magic', as it's called, 'will answer your need, but not to your demand; it will respond to your heart, but not to your head': we have little choice in the matter. If our technology, our new magic, fails to recognise this old magic of the earth, it will, in the long term, be heading for a collapse which will be anything but amusing. Without that wry sense of humour to help us interpret its strange jokes, nature has the last laugh on us every time.

What's the use?

It's easy to become fatalistic about the current state of technology. The great promises of each new technology – the nuclear technology of thirty years ago, the micro-processor revolution of the last decade – seem to have been lost in meeting up with reality. Nothing is as easy as it seems, it's no fun any more, people say; the joy and the magic have left it.

So it's left to science to pick up the credit for 'progress', and for commerce to make money now and leave the problems of tidying up the mess for someone else to solve. The more glamorous and exciting the technology, the less anyone wants to work at being dustman; the big toys of the nuclear technologies, for example, have given us problems that can, we're told, only be solved by yet more expense on bigger technological toys, that will themselves create even greater problems that can only be solved by even bigger technological toys... And so it goes on. 'Toys for the boys'; while the real dangers for our future are swept beneath an imaginary carpet, shuffled into invisibility behind a dazzling display of pyrotechnics and a neat 'somebody else's problem' field (to use Douglas Adams' depressingly-accurate term).

> "The only difference between men and boys
> is the ever-increasing size and cost of their toys"
>
> (American proverb)

As we've seen, much the same is true in medicine. Bigger and more complex technological toys, to deal with wayward lumps of meat in a scientific manner. It works, after a fashion, but there's not much magic, and very little respect for people as people.

And none of it is efficient in a real sense; it's not exactly reliable, it's hardly elegant, and it rarely seems to be appropriate.

But *we* are technology: we *are* the magic. We cannot hide from it in some golden-age, 'New-Age' Avalonian mists: we are *here*, *now*. It's *our* problem.

It's generally easier, too, to start from where we are rather than from where we are not.

We cannot start as idealists: our nature *does* include greed, laziness, avarice and the rest of the biblical deadly sins; so any technology, any way of working on the world, is unlikely to work if it does not allow for these aspects of reality.

We cannot start as scientists, either: that would allow us, perhaps, to start from where we are, but would only allow us to move as far as the limitations of logic will let us – and that's not very far in the messy reality of the real world.

We can only really start, perhaps, from *who* we are. Not defined in terms of what we do, but more described in terms of where we 'be'. In a context in which 'anything goes', and in which (through that paradox of 'things have not only to be seen to be believed, but also to be believed to be seen') we play a direct part, our approach to reality, and working on that reality, is a distinct and distinctive part of that reality. Beyond the normal 'default reality' of physics, and the not-so-normal games of the natural world, *we* decide what is real and what is not: for others as well as for ourselves.

The world and its effects are of *our* choosing. If we want to change it... ...we can change it. By looking at ourselves, through the technology that is the expression of ourselves.

We are all magicians, whether we like it or not: we play a direct part in the reality of the world that we and everything else around us will experience.

And as magicians, isn't it perhaps about time we got round to being good at it?

Bibliography

The books listed here were either referred to specifically in the text, or are the books from which specific references were taken. It is not, of course, a complete bibliography!

Douglas Adams, *The Hitchhiker's Guide to the Galaxy*, Pan, 1976
Anon., *SSOTBME – an essay on magic, its foundations, development and place in modern life*, The Mouse That Spins, 1976
Kenneth Batcheldor and Colin Brookes-Smith, *Manual of Advanced Psychokinetic Procedures*, Society for Psychical Research, 1970
W.I.B. Beveridge, *The Art of Scientific Investigation*, Heinemann, 1957
Edward de Bono, *Practical Thinking*, Jonathan Cape, 1971
James Burke, *The Day the Universe Changed*, BBC Publications, 1985
Charles Fort, *The Book of the Damned*, Abacus, 1973 (original U.S. publication 1919)
Tom Graves, *Dowsing: Techniques and Applications*, Turnstone, 1976
Tom Graves, *Needles of Stone*, Turnstone, 1978
Philip Harben, *The Grammar of Cookery*, Penguin, 1965
Eugen Herrigel, *Zen and the Art of Archery*, 1953
Thomas Kuhn, *The Structure of Scientific Revolutions*, University of Chicago, 1970

Lao Tzu, *Tao Te Ching*, trans. Gia Fu Feng and Jane English, Wildwood, 1976

James Lovelock, *Gaia: a new look at life on Earth*, Oxford University Press, 1979

John Michell and Bob Rickard, *Phenomena*, Thames and Hudson, 1977

Robert M. Pirsig, *Zen and the Art of Motorcycle Maintenance*, Bodley Head, 1979

Idries Shah, *The Exploits of the Incomparable Mulla Nasrudin*, Jonathan Cape, 1966

M.W. Thring and E.R. Laithwaite, *How to Invent*, Macmillan, 1977